THE OPE[N]
MATHEM[A]
AN INTER[...]
MA290: T[O...]   [S]

BLOCK 2   FROM T[H...]
SEVENTE[...]

UNIT 8

DESCARTES: ALGEBRA AND GEOMETRY

PREPARED BY JEREMY GRAY FOR THE COURSE TEAM

THE OPEN UNIVERSITY

CONTENTS

This unit forms part of an Open University course. The set book for the course, to which reference is made as **SB**, is:

John Fauvel and Jeremy Gray (editors), *History of Mathematics: A Reader*, Macmillan 1987.

Acknowledgements

Grateful acknowledgement is made to the following sources for material used in this unit: *Figure 1*, Royal College of Physicians; *Figures 2* and *7*, The Science Museum.

The Open University, Walton Hall, Milton Keynes.

First published 1987. Reprinted 1989, 1996.

Designed by the Graphic Design Group of the Open University.

Typeset in Great Britain by Santype International Ltd, Salisbury.

Printed in Great Britain by BPC Wheatons Ltd, Exeter.

ISBN 0 335 14252 4

This text forms part of the correspondence element of an Open University Second Level Course.

For general availability of supporting material referred to in this text, please write to Open University Educational Enterprises Limited, 12 Cofferidge Close, Stony Stratford, Milton Keynes, MK11 1BY, Great Britain.

Further information on Open University courses may be obtained from The Admissions Office, The Open University, P.O. Box 48, Milton Keynes, MK7 6AB.

1.3

8.0 INTRODUCTION

In this unit, we concentrate chiefly on the mathematical work of René Descartes (1596–1650). His *La Geometrie*, published as an appendix to a philosophical treatise in 1637, was not just a new and decisive way to introduce algebra into geometry; it was to transform ways of thinking about geometry altogether. In Section 1 we look at the context within which he worked, and focus on two key words: *algebra*, and *analysis*. Algebra was originally introduced as a means to analyse a geometrical problem, i.e. to break the problem down into manageable pieces; and indeed Descartes was excited by the idea that the force of mathematical analysis could also be made to enrich philosophical analysis. In Section 2 we look at how he came to write *La Geometrie*, and briefly at what it says about how to solve mathematical problems. One particular problem is considered in more detail in Section 3, the so-called *Pappus problem*. This played a crucial role in showing how algebraic equations can be used to describe curves. In Section 4 we look at how Descartes' ideas were received, and in Section 5 at how Isaac Newton, in particular, responded to them. We shall see that Newton had rich and complicated views about algebra and geometry, and about how the two should be related, which give a fascinating insight into how he perceived his Cartesian inheritance.

8.1 ALGEBRA AND ANALYSIS

In *Unit 7* you met some of the sprawling mathematical achievements of the early seventeenth century, and saw that many of its most profound insights were destined, for various reasons, not to be pursued for a hundred years or more. In this unit, in contrast, we follow the story of a development that went from strength to strength almost from its inception: the introduction of algebraic methods. The transformation of mathematics thus wrought was immense. In the course of the seventeenth century mathematics acquired a great many new objects to study, problems to deal with, and methods for tackling them. Moreover, mathematicians were emancipated from their Greek inheritance. Whereas Viète and even Fermat thought of themselves as rediscoverers of lost Greek knowledge, by the end of the century no-one felt obliged to make that comparison any more. They were by then their own masters.

The historian Michael Mahoney has called the transition from the geometric to the algebraic mode of thought 'the most important and basic achievement of mathematics at the time'. For the century which saw the invention of the calculus, this is a bold claim. Later you will be in a position to assess the importance of the algebraic revolution for yourself, but first we seek to understand it better by asking: why did it happen? What made possible so fertile a transformation of mathematical practice?

M. Mahoney, 'The beginnings of algebraic thought in the seventeenth century', in Stephen Gaukroger (ed.) *Descartes: Philosophy, Mathematics and Physics* (Harvester Press, 1980) p. 141.

One way to answer the question is to invoke the *power* of the new methods, and to argue that it was the availability of new techniques which led directly to the new conquests. Plainly, in a mathematical culture intent on finding new results, a new technique might well drive things a long way. The new methods derived partly from the analytic art of Viète. Recall from *Unit 5* that Viète's idea was to use algebra to analyse a problem, in the sense that Pappus used the term *analysis*. That is, you assume that the solution to your problem is known, and then argue until you reach a known truth. This process is followed by a synthesis, using rigorous deduction from the known truth until the desired solution is validly demonstrated. Viète thought of his work as supplementing an essentially geometrical and Greek way of proceeding; his original contribution, in his own opinion, was in 'exegetics', the solution of the equations produced by the analysis. As the seventeenth century proceeded, the tendency was to grant algebraic analysis a rigour of its own and to drop the synthesis. Thus a radically new technique and mode of mathematical justification evolved out of interpreting the Greek tradition, a tradition whose emphasis (on synthetic proof) ran in the opposite direction. This is too major a shift, Mahoney has argued, to be accounted for by the efficacy of the new methods alone; other factors must have disposed people to accept the new algebraic approach. Mahoney draws attention to two features of the broader intellectual concerns of the period that may have conditioned the new conception of mathematics: the search for a universal language, and the pedagogic tradition.

Many scholars throughout the seventeenth century (as you saw in *Unit 7*, Section 5) were pursuing the quest for a natural, universal symbolism to reveal truths about the world. Algebraic language came to be seen as a paradigm of successful symbolising which could have wider application. The philosopher John Locke certainly saw a connection. He wrote in 1690:

> They that are ignorant of *Algebra* cannot imagine the Wonders in this kind are to be done by it: and what farther Improvements and Helps, advantageous to other parts of Knowledge, the sagacious Mind of Man may yet find out, 'tis not easy to determine . . .

> Mathematicians abstracting their Thoughts from Names, and accustoming themselves to set before their Minds, the *Ideas* themselves . . . have avoided thereby a great part of that perplexity, puddering, and confusion, which has so much hindred Mens progress in other parts of Knowledge.

> And who knows what Methods, to enlarge our Knowledge in other parts of Science, may hereafter be invented, answering that of *Algebra* in Mathematicks, which so readily finds out *Ideas* of Quantities to measure others by, whose Equality or Proportion we could otherwise very hardly, or perhaps, never come to know?

John Locke, *An Essay Concerning Human Understanding*, Book IV (1690) Chapters III, XII.

So there were wider intellectual currents of the period encouraging the development of symbolic algebra. Locke was writing late in the century, but earlier writers too—such as Descartes—discussed explicitly the power and value of symbolic expression.

The other intellectual concern singled out by Mahoney was a changing conception of mathematical pedagogy. This stemmed in particular from the influence of the sixteenth-century teacher and textbook writer Peter Ramus. Ramus had castigated Euclid on the very grounds that others had praised him, namely for the *Elements'* rigour and deductive logical structure. In Ramus' view this style was a fault, for rigorous proofs conveyed neither clarity nor insight. Plainly, this is a debate about the teaching of mathematics that is still with us! But it is more than that, for it was a constant source of complaint in the early seventeenth century that the classical texts were so opaque. The mathematical results were both hard to learn, and impossible to extend. There was a pressing need for another mathematical way to be found, and by the force of his example Ramus helped legitimise arguments which made up in intelligibility what they lacked in rigour.

Also known as Pierre de la Ramée, 1515–1572.

A powerful demonstration of the new algebra arising in the service of a new teaching approach was the influential textbook of William Oughtred, *Clavis Mathematicae* of 1631. In the preface to the first English translation (1647) Oughtred explained his intentions. Please *read that now* (**SB** 9.F1(a)).

Question 1 The seventeenth century was characterised above as one during which mathematicians were emancipated from their Greek inheritance. How far along this road of emancipation do you judge Oughtred to have travelled?

Comment ────────────────────────────────

In some ways he has departed radically from classical practice. His book was not written synthetically, and with words; but analytically, and with symbols. So Oughtred had abandoned the classical proof structure as a way of presenting results, because he felt (as Ramus had done earlier) that this did not lead to understanding.

But the purpose of the book ('my scope and intent') was to lead to understanding of the classical authors, Euclid, Archimedes, Apollonius etc. It was not to investigate contemporary mathematics nor to aid current researches. (It did have that effect later, as it happens, but it is interesting that Oughtred did not have that intention.)

So Oughtred can be seen to have broken with the Greek mathematical tradition, both in proof structure and in his wholesale symbolisation—for the purpose of understanding that tradition better. This seems at once a noble and a perverse endeavour; emancipation had further to go. ■

Debates about the relative virtues of analysis and synthesis continued, both in mathematics and in the broader intellectual context, mentioned earlier, of what method should be followed for the discovery and teaching of truth more generally. Descartes was particularly concerned with such matters. The following comparison of analysis and synthesis was part of his reply to critics of his philosophical work, *Meditations*. In reading this try to ascertain Descartes' reasons for preferring the method of analysis.

> Analysis reveals the true way in which a thing was found methodically and, as it were, *a priori*, so that, if the reader wishes to follow it and pay sufficient attention to everything, he will understand the matter no less perfectly and make it no less his own than if he himself had found it. But it has nothing by which to incite belief in the less attentive or hostile reader. For if he should not perceive the very least thing brought forward, the necessity of its conclusions will not be clear; often it scarcely touches on many things which should be especially noted, because they are clear to the sufficiently attentive reader.
>
> Conversely, synthesis clearly demonstrates, in a way opposite to analysis and, as it were, *a posteriori* (even though the proof itself is often more *a priori* in the former than in the latter), what has been concluded, and it uses a long series of definitions, postulates, axioms, theorems, and problems, so that, if one of the consequents is denied, it may at once be shown to be contained in the antecedents. Thus it forces assent from the reader, however hostile or stubborn. But it is not as satisfying as analysis; it does not content the minds of those wanting to learn, because it does not teach the manner in which the thing was found.
>
> The ancient mathematicians used to employ only synthesis in their writings, not because they were simply ignorant of the other, but, as I see it, because they made so much of it that they reserved it as a secret for themselves alone.
>
> In fact, I have followed in my *Meditations* only analysis, which is the true and best way of teaching.

Second Reply to Objections against the *Meditations*, 1641, cited in Mahoney, *Descartes*, p. 149.

So analysis is the best way to teach results because it shows you how they are found, and so it leads to understanding in those who want to learn. Synthesis forces you to acknowledge that a result is correct by showing you that it must be true.

From this and the earlier quotations it does seem that it was the power of algebraic analysis as a discovery method that made the introduction of algebra important. We shall conclude our stage-setting, then, by turning from *analysis* to look at *algebra*. What could it be about algebra that made it so attractive?

Mahoney gives a three-point summary of what came to characterise algebra during the seventeenth century:

(i) It is symbolic, and it symbolises not only objects (like lines) but also operations (like addition) and relations (like equality).

(ii) What the objects *are* becomes less important. Algebra is ontologically uncommitted, in the sense that you need not know what x means, as long as you know the rules for handling such symbols.

(iii) The operations and relations become the focus of attention, rather than the objects. The rules for handling symbols become of prime significance.

Viète's importance to us, you saw in *Unit 5*, rests on his giving to algebra the generality of geometry. The symbols, denoting variable magnitudes, acquired in his hands a logic of their own, the logic of algebra rather than geometry. A geometric argument is valid because its statements accord with the nature of the magnitudes under discussion. An algebraic argument, however, is valid because its almost meaningless symbols have been validly manipulated. This distinction, hard though it is to make, and obscure though it can be in the sources, should at least indicate one thing: the algebraic method, being more formal, is more general. To help us focus on the distinction, we conclude this section by looking at an example of algebraic analysis which exemplifies many of the processes we have been thinking about: Fermat's study of curves in 1636.

Fermat was among the first to extend to curves the approach which Viète had taken towards general magnitudes, namely a conscious analytical exploration of the connection between problems and their solutions. We know a little about how Fermat came to work in this way. It seems that his earliest mathematical contact, in the late 1620s, was with one Etienne d'Espagnet who was in possession of some of Viète's unpublished manuscripts. D'Espagnet communicated these to Fermat, who proceeded to study them avidly. It is not surprising, therefore, that Fermat's first interest in geometry was an attempt to restore a lost book of Apollonius (called *On Plane Loci*), for Viète too had engaged in seeking to restore lost texts. This attempt seems to have suggested to Fermat that the preparatory algebraic analysis he carried out was capable of forming a more general method than the classical one, and in 1636 he wrote a paper describing his own method (*An introduction to plane and solid loci*). However, he did not publish it. It was first published in 1679, after his death; coming so long after the publication of Descartes' work, its influence on the history of mathematics was slight.

Fermat used Viète's notation for constants, variables, and equations, and set himself the task of analysing equations of the first and second degrees in two unknowns. He found that any such equation could be reduced to one of seven types, which he interpreted geometrically as various kinds of conic section. This was one of the first occasions when algebraic analysis was treated not simply as a means to an end, but as an end in itself. An algebraic object (an equation) was both interpreted geometrically and taken as the main object under investigation. Thus, when Fermat considered (in his Viètan notation)

$$A \text{ in } E \text{ aeq } Z \text{ pl,}$$

an equation between variable magnitudes A and E and a constant magnitude Z, he showed that the corresponding locus was a hyperbola. What Fermat had done was to show that for the investigation of conics, an analysis by means of simple equations is a convenient and productive method.

We can modernise this as $AE = Z^2$, since Fermat wrote 'in' for 'times', 'aeq' for 'equals', 'pl' for 'plane': we get the more familiar equation $xy = c^2$ on switching to 'our' letters.

Question 2   Look back over this account of Fermat's investigations. In what way does it exemplify what we have been saying about algebra?

Comment ────────────────────────

It is symbolic, and has its own logic, that of algebra. Rules of inference are rules about manipulating symbols. It is ontologically uncommitted, in the sense that A and E are perfectly general magnitudes until he chooses to pin them down as relating to points on a hyperbola. But notice that it is still subordinate to the Greek legacy. Fermat was offering a way into Apollonius' theory of conics, and not, it would seem, a way beyond it.   ∎

8.2   DESCARTES

It would scarcely be an exaggeration to say that, as Galileo is to physics and Kepler to astronomy, so is Descartes to mathematics: the modern study of each subject started with these men. This was the opinion of many mathematicians of the next century, and later. For example, John Stuart Mill hailed Descartes' work as 'the greatest single step ever made in the history of the exact sciences'. In fact, most of Descartes' work was in natural science or mathematics, although he is perhaps best known today for his philosophical work. When Descartes was writing, the new science had scarcely established itself as a method of inquiry and it had few triumphs to its name. Opposition to it came largely from philosophical sceptics, who disputed the claim that the use of this novel scientific method could add to knowledge. There was also opposition, to a lesser extent, from the philosophical and

An Examination of Sir W. Hamilton's Philosophy (1878) p. 617, cited in Jonathan Rée, *Descartes* (Allen Lane, 1974) p. 28.

theoretical orthodoxy which accepted traditional Aristotelian physics. The over-riding purpose of Descartes' work was to establish that rational inquiry could lead to knowledge, and his mathematical work was, in part, intended to show how it could be done. Indeed, as we shall see, mathematics was for Descartes the paradigm case of rational inquiry, but he did not disdain science. He went on to investigate a theory of optics, and to outline a theory of planetary motion which Newtonian gravitational theory was not to supplant, in France at least, until well into the eighteenth century.

In his *Discourse on Method* (1637), Descartes gave an account of how he came to his views about the role of mathematics in the pursuit of true knowledge. One day in the winter of 1619, he claimed that 'I spent the whole day shut up alone in a stove-heated room', and simply thought. He came to feel that 'buildings conceived and completed by a single architect are usually more beautiful and better planned than those remodeled by several persons using ancient walls . . .'. In short, he felt that it would be necessary to carry out a programme for remodelling knowledge completely by himself—an impressive ambition, even for a young man of twenty-three. It followed, he realised, that he had to doubt every idea which he might have acquired from others in his childhood. That night (it was November 10th, 1619) he had three dreams, according to his first biographer Adrien Baillet, which reinforced him in his sense of mission. Thus began the intellectual odyssey which was to take nearly twenty years before reaching its first published fruits, the *Discours de la Méthode* with its celebrated appendix *La Geometrie*, in 1637.

Descartes, *Discourse on Method*, ed. J. L. Lafleur (Bobbs-Merrill, 1964) p. 10.

More prosaically, the chief mathematical and scientific influence on Descartes seems to have been Isaac Beeckman, whom Descartes met in Breda in 1618, while serving as a gentleman volunteer in the army of Prince Maurice of Nassau. He wrote to Beeckman on 26th March 1619, outlining his ambitions in mathematics.

Question 3   In the following extract from Descartes' letter to Beeckman, what sort of approach was he advocating to geometrical problems?

> I hope to prove . . . that certain problems can be solved with straight and circular lines only; that others can only be solved with other curved lines which originate in one single motion, and which therefore can be traced by the new compasses, which I do not think are less certain and geometrical than the ordinary ones with which circles are drawn; and that finally other problems can be solved only by curved lines originating from different motions that are not subordinate to each other and that certainly are only imaginary (*imaginariae*); such a curve is the well-known quadratrix. I think that one cannot imagine problems that cannot be solved by at least these lines; but I hope to be able to demonstrate which questions can be solved by the first or the second method and not by the third; so that in geometry nothing remains to be found.

Isaac Beeckman (1588–1637) was a Dutch scholar, craftsman and teacher. The Netherlands were a centre of Ramist thought (views on education influenced by Ramus) at this time.

Descartes literally had in mind a machine made up of hinged rods which would draw the curves in exactly the same sense that a compass draws a circle.

Quoted in H. J. M. Bos, 'On the representation of curves in Descartes' Geometrie', *Archive for History of Exact Sciences*, **24** (1981) pp. 326–7.

Comment ───────────────────────────────
This reads like a version of the classification of geometric problems given by Pappus (*Unit 3*, Section 3). But the greater instrumental bias of Descartes' views is noticeable, and also the confidence with which he feels that once geometric problems are correctly allocated to their appropriate construction, all geometry would have been sorted out. And there is a hint in the final sentence that Descartes believed the problems of his third category to be outside geometry altogether. ■

We shall discuss in Section 3 what problems lurk in Descartes' ideas about curves, but this is certainly a very optimistic programme, strikingly beyond his mathematical abilities to carry out at the time. We shall also see that his view that some problems are unsolvable geometrically was, by contrast, unduly pessimistic.

Descartes studied mathematics intensely after 1619, attempting in particular to devise a clear and simple algebraic symbolism for analysing mathematical problems. Viète had had the same aim, but it seems that Descartes' later claim was true, namely that he did not know of Viète's work. In 1631, having by now settled in Holland in order to enjoy the benefits of religious tolerance, he was introduced by a friend to a problem discussed in the *Mathematical Collection* of Pappus, the 'locus to 3 or 4 lines'. All this time he had tried, with little success, to extract from mathematics the key to correct reasoning. Descartes' resolution of the problem of

Figure 1   René Descartes (1596–1650)

Pappus (which we look at in more detail in the next section) was to be his sign that he was finally achieving his goal.

Descartes first published his method of rational enquiry in his *Discourse on Method* (1637). This, however, scarcely discusses the method itself but moves on rapidly to outline the philosophy Descartes had by then developed, and to show off other of its fruits in three further essays printed at the end of the *Discourse*. The first of these, on optics, contains the first published statement of the sine law of refraction, and an analysis of lenses and the eye. The second appendix, on meteorology, contains an explanation of primary and secondary rainbows. The third, with which we are chiefly concerned, is the justly celebrated *La Geometrie*. Before turning to this work and examining it in detail, let us see how Descartes discussed, in the *Discourse*, the role of mathematics. Please *read now* **SB** 11.A1.

The full title is instructive: *Discourse on the Method of rightly conducting the Reason in the search for Truth in the Sciences.*

Question 4   On the basis of paragraph 3 of **SB** 11.A1, try to decide what lesson Descartes drew from mathematics. What did he intend his contribution to mathematics to be?

Comment ————————————————————————————

Descartes drew from mathematics the idea that all knowledge could be presented in deductive chains, starting from simple truths and using only sound reasoning. He would not learn all mathematics, but, starting from conclusions about straight lines, he would bring together geometry and algebra, thereby improving them both.   ■

Descartes did not consider himself as just a mathematician, but as a philosopher in the broadest sense of the term. In the *Discourse* he described his aim as 'seeking the true method of obtaining knowledge of everything which my mind was capable of understanding'. Striking though the word *everything* is in the present context, the key word here is *method*. As we saw, it is the search for a method that brought Descartes to mathematics. He also tells us that when he turned to mathematics he found a geometry that was fatiguing and an algebra that was obscure, and so he set out to devise his own rules for discovering the truth.

The preoccupation with *method* was not unique to Descartes. It was a concern shared by many, and was influenced by the spread of Ramus' ideas in particular.

Question 5   Please look now at the rest of **SB** 11.A1, especially the four rules in paragraph 2. Think about their implications for the study of mathematics, noting especially the second rule. See if you can decide in which way the whole extract has gone beyond what Descartes said to Beeckman in 1619 (quoted in Question 3).

Comment ————————————————————————————

I was struck by how much the second rule looks like analysis (in the sense that Pappus and Viète used the term), and how Descartes ended paragraph 3 by talking about using 'numbers' to express relations between lines. That is, analysis of a problem was to be expressed in algebraic language. This singling out of algebra as the way to analyse a problem is the crucial step taking him beyond his earlier position.   ■

Later generations were not to find it easy to say when something was 'certainly and evidently' true, but nor were they to propose any better characterisation of statements which could be taken as bed-rock in an intellectual enquiry. One may suppose that Descartes regarded his four rules rather as brief notes than as definitive statements—indeed, he said as much in a note he sent to his old schoolmaster, to whom he confided that the *Discourse* did not teach but only functioned as a notice for the method. They encapsulated a philosophy he thought worth proceeding with, so they are, perhaps, pieces of 'good advice'—but Descartes took them very seriously.

We now turn to *La Geometrie* to see how successfully Descartes fulfilled his promises. Although Descartes would have agreed with all of his contemporaries that geometry is the study of magnitudes, which can conveniently be represented by lines, he introduced a crucial simplification almost at once. Whereas others viewed the product of two lengths as an area, and the product of three lengths as a volume, Descartes explained how magnitudes can be manipulated geometrically and the results described algebraically.

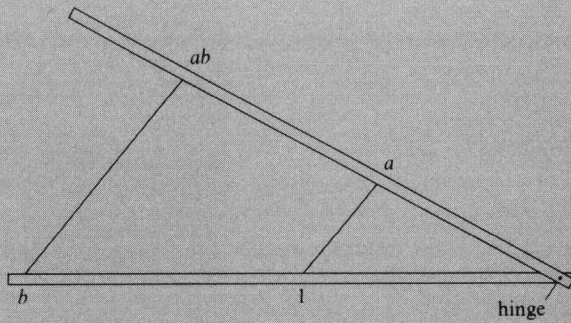

Box 1 The sector

Descartes' approach to multiplication is exemplified in a mechanical instrument widely available in his day, the *sector*. This device, which may have been invented by Galileo among others, consisted of two hinged rods upon which scales were marked. To multiply two numbers a and b with it, set a pointer on one scale at a, another pointer on the other scale at b, and slide a pair of parallel lines into position as shown. In short, one multiplied numbers exactly as Descartes multiplied magnitudes. Descartes saw in this argument a conclusion which had escaped a generation of practitioners.

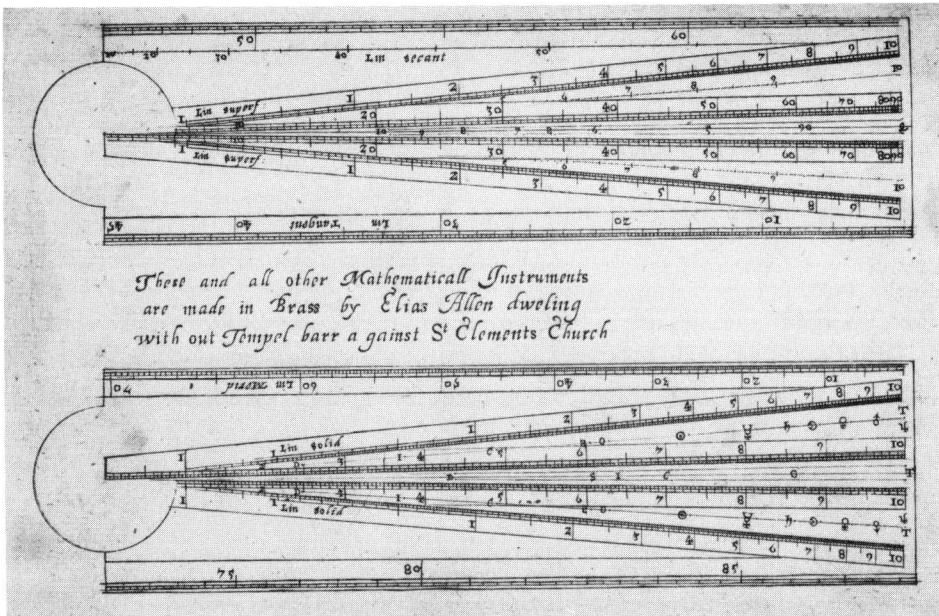

Figure 2 An example of a sector designed by E. Gunter, from the 1636 edition of his book, *The Description and Use of the Sector*

In **SB** 11.A2 he describes examples of multiplication and of division; see if you can spot the crucial new point of view that Descartes introduced.

Taking multiplication as an example, we can argue the following. To multiply BD by BC, Descartes took BA as a unit segment, and drew the segments on different lines through the vertex B (see Figure 3). He formed two triangles, ACB and DEB, choosing E so that DE and AC were parallel. Since AC and DE are parallel, these triangles are similar, so

$$\frac{BC}{BA} = \frac{BE}{BD} \text{ and hence } BC \cdot BD = BE \cdot BA;$$

but BA is of unit length so BE is equal to the product $BC \cdot BD$, as claimed.

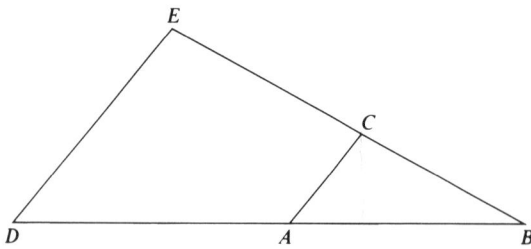

Figure 3

The remarkable thing is that Descartes took the length *BE* to be the answer, whereas the whole of classical antiquity took a product of lengths to be an area. By deliberately suppressing the fact that *BA* was a unit of *length*, Descartes introduced dimensionless quantities into mathematics (or rather, all quantities became one-dimensional)—this is the new point of view referred to above. Previously geometry had dealt with lengths, whose products were areas, and instead of division had used the concept of ratio. Now it could deal entirely with lengths, whose products are again lengths, and whose quotients are also lengths. In this respect it is like elementary algebra, which deals with numbers; and the unification of algebra and geometry was to be vital to Descartes' programme. But it is unlike Viète's algebra, which was as conscious of dimension as was Greek geometry. Viète was developing his vision of the true, classical mathematics; Descartes was making a fresh start. From Descartes' letter to Desargues, you will know how complicated Descartes found the classical arguments about ratio and proportion (especially in Desargues' hands). Here he is presenting his streamlined version, in which quantities behave like numbers in that their products and quotients are of the same kind; we know that it was a successful move, because it is the style we all write in today.

SB 11.D4 (Question 8 of *Unit 7*)

Descartes was not the first to propose such a simplification. What was crucial about his proposal was the extent to which he spelled out its implications. For someone who claimed to expect no advantage from his study of mathematics 'except to accustom my mind to work with truths', he went a long way to reformulating the very subject itself.

SB 11.A1

The same spirit of simplification can again be seen in the final paragraph of this extract, the construction of a square root. Its significance lies not in the construction itself, which was the standard classical one for finding a mean proportional, but in the way Descartes has discarded the classical constraints of proportional terminology, treating all lines as representing comparable magnitudes regardless of the operations producing them.

Euclid's *Elements* VI, 13; in **SB** 3.C4

To this conceptual simplification, Descartes added the notational simplification of representing lines by single, lower-case, letters (see Figure 4). This made possible a

LIVRE PREMIER. 299

gnes ſur le papier, & il ſuffiſt de les deſigner par quelques lettres, chaſcune par vne ſeule. Comme pour adiouſter la ligne B D a G H, ie nomme l'vne *a* & l'autre *b*, & eſcris *a* + *b*; Et *a* -- *b*, pour ſouſtraire *b* d'*a*; Et *ab*, pour les multiplier l'vne par l'autre; Et $\frac{a}{b}$, pour diuiſer *a* par *b*; Et *aa*, ou a^2, pour multiplier *a* par ſoy meſme; Et a^3, pour le multiplier encore vne fois par *a*, & ainſi a l'infini; Et $\sqrt{a^2 + b^2}$, pour tirer la racine quarrée d'$a^2 + b^2$; Et $\sqrt{C.\, a^3 - b^3 + abb}$, pour tirer la racine cubique d'$a^3 - b^3 + abb$, & ainſi des autres.

vſer de chiffres en Geometrie.

Où il eſt a remarquer que par a^2 ou b^3 ou ſemblables, ie ne conçoy ordinairement que des lignes toutes ſimples, encore que pour me ſeruir des noms vſités en l'Algebre, ie les nomme des quarrés ou des cubes, &c.

Il eſt auſſy a remarquer que toutes les parties d'vne meſme ligne, ſe doiuent ordinairement exprimer par autant de dimenſions l'vne que l'autre, lorſque l'vnité n'eſt point determinée en la queſtion, comme icy a^3 en contient autant qu'*abb* ou b^3 dont ſe compoſe la ligne que i'ay nommée $\sqrt{C.\, a^3 - b^3 + abb}$: mais que ce n'eſt pas de meſme lorſque l'vnité eſt determinée, a cauſe qu'elle peut eſtre ſouſentendue par tout ou il y a trop ou trop peu de dimenſions: comme s'il faut tirer la racine cubique de *aabb* -- *b*, il faut penſer que la quantité *aabb* eſt diuiſée vne fois par l'vnité, & que l'autre quantité *b* eſt multipliée deux fois par la meſme.

P p 2 Au

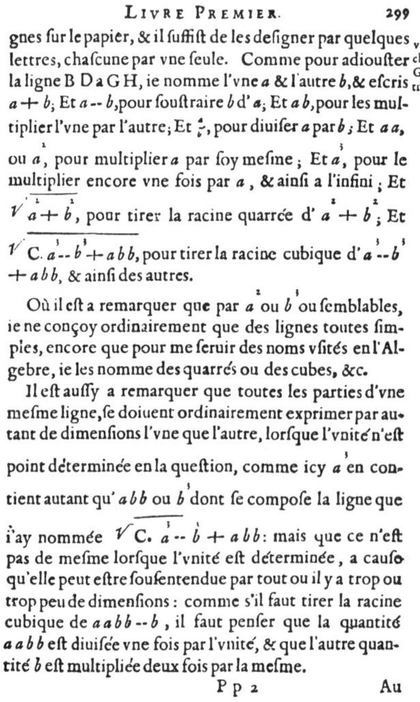

Figure 4 Descartes, *La Geometrie* (1637). The first full sentence says, in part, 'Thus to add the line BD to GH, I call the one *a* and the other *b*, and write $a + b$; . . . ; and *aa*, or a^2, to multiply *a* by itself; and a^3, to multiply it once again by *a*, and thus to infinity; . . .'

much more succinct handling of the algebraic formulae as the underlying geometrical constructions grew more complex. Descartes was now ready to embark on the detailed description of his general programme, which forms the next extract you should read. It is quite long, and difficult in places—Descartes seems to have been regarded as an arrogantly casual expositor by his successors. Read it over once to appreciate its structure, then again to see what is going on in each paragraph, before reading the remarks below. Please *turn to* **SB** 11.A3 *now*.

The passage opens with the claim that a method will be developed for solving any problem, and this method is to involve equations. The argument proceeds at this high level of generality until we are told (in [4]) that the aim is to obtain a single equation by eliminating all but one of the unknowns. It then descends, by means of the lofty remarks in [5] and [6], to an example (in [7]), the quadratic equation $z^2 = az + b^2$.

Let us now turn to the details. Paragraph [1] is plainly a process of analysis: 'we first suppose the solution already effected', and he gives names to everything in sight. We then unravel the problem by manipulating the lines until we have obtained equations between them. That enables the elimination of quantities until, one hopes, a single equation is left—an outcome assumed rather than explicitly justified here. In [2], Descartes extended the method to encompass problems involving many unknowns, and in [3] he displayed the sort of equation he expected to get *en route* to the solution. Paragraph [4] will detain us later, when we unpack what Descartes meant by a problem 'constructed by means of circles and straight lines'. Just note that it does not mean that the answer *is* a circle or a straight line. Indeed, the answer is most likely to be a number or a line-segment representing a magnitude.

What can be said about Descartes' generous exercise of self-restraint in [5], his unwillingness to deprive the reader of satisfaction and educational benefit? Descartes' claim that there was 'nothing here so difficult' was perhaps disingenuous. The English translators of *La Geometrie* quoted from a letter which Descartes wrote to Mersenne in 1637:

> As to the suggestion that what I have written could easily have been gotten from Viète, the very fact that my treatise is hard to understand is due to my attempt to put nothing in it that I believed to be known either by him or by anyone else . . . I begin where he left off.

D. E. Smith and M. L. Latham, tr. *The Geometry of René Descartes* (Open Court, 1925) p. 10, note 17.

The example in [7] is probably the best way in. First, the details. Descartes put $LM = b$, $NL = \frac{1}{2}a = NP = NO$, where a and b are given. By Pythagoras' Theorem

$$MN^2 = LN^2 + LM^2 = \tfrac{1}{4}a^2 + b^2,$$

so $$MN = \sqrt{(\tfrac{1}{4}a^2 + b^2)};$$

thus $$OM = ON + NM = \tfrac{1}{2}a + \sqrt{(\tfrac{1}{4}a^2 + b^2)}.$$

The marvellous thing is that $OM = z$ is a solution of $z^2 = az + b^2$, and this solution can be obtained entirely algebraically by completing the square. So paragraph [7] shows how, in a simple case, one can proceed by a geometric construction involving circles and lines to solve a quadratic equation. By implication, we suppose that Descartes had other constructions for more complicated equations; how else could he claim to be able to solve *any* problem?

So what have we learned? That Descartes laid claim to a general method to solve problems by algebra, and that the solution of the equations which arise can be effected by geometrical constructions. What we have not seen is the method at work, actually obtaining the equations, nor have hard equations been solved. That was to follow, and forms the subject of our next section. You will see that Descartes took the Pappus problem, which he was so happy to have solved four years earlier, and worked it through—first for three or four lines, and then for any number. He was clearly impressed with how completely he had transcended the classical writers, and so too were his readers. That is the great force of his work.

8.3 LOCUS PROBLEMS

The *Mathematical Collection* of the late Hellenistic commentator Pappus of Alexandria had been known to the European mathematical community since 1566, when Federigo Commandino included extracts in his edition of Apollonius' *Conics*. It had been studied sporadically for over half a century when Descartes' friend Jacob Golius drew his attention, in 1631, to a problem Pappus had discussed in connection with the *Conics* of Apollonius. Descartes' success in solving and developing his ideas around this problem was to become central to the exposition of *La Geometrie*.

Unit 5, Section 5

Before we explore the details of the problem, and Descartes' solution and further developments, it will be helpful to see just what sort of a problem it was. It was a *locus problem*, which is a variant on the long-standing Greek preoccupation with defining curves by property. It is the converse of being given a curve and seeking to find the property common to its points—in a locus problem the property is given (expressed in terms of lines and distances, say) and one wants to find all the points with that property. Thus the locus of all points which have the property of being a particular distance from a given fixed point is a well-known curve, the circle. For another example, what is the locus of all points equidistant from two fixed lines (where the distance from a point to a line is measured along the perpendicular to the line)? Here the locus is a straight line, and in fact is the line bisecting the angle made by the fixed lines: any point on the line of bisection has the property that it is equidistant from the two lines initially given (Figure 5).

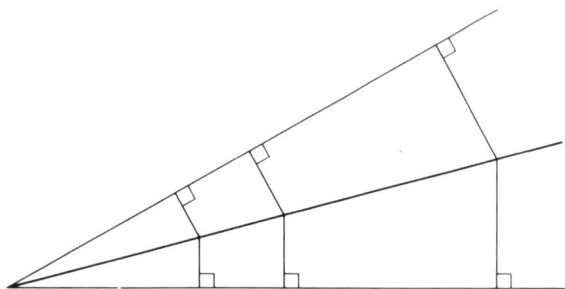

Figure 5

The problem to three or four lines, known as Pappus' problem, is a generalisation of this last problem. Consult Box 2, which explains the problem, before reading Pappus' account.

Box 2   The locus to three or four lines

The problem for the locus to four lines is described below.

These ingredients are specified in advance:

(i)   four straight lines $(\ell_1, \ell_2, \ell_3, \ell_4)$;

(ii)   four angles $(\alpha_1, \alpha_2, \alpha_3, \alpha_4)$;

(iii)   a ratio (k).

Suppose the lines lie like this:

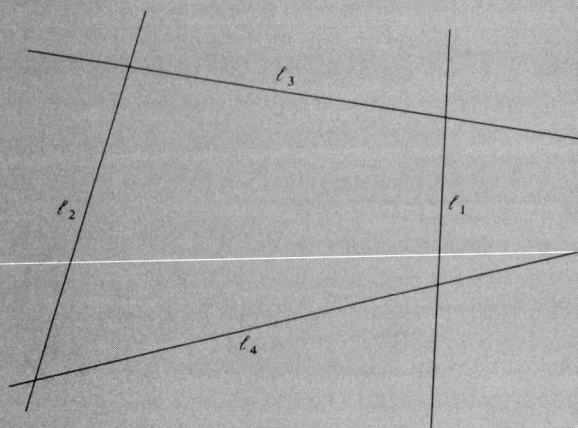

Then you seek the positions of a point P whose distances p_1, p_2, p_3, p_4 from the given lines have the property $p_1 \cdot p_2 = kp_3 \cdot p_4$, *where the distances are measured along lines from P meeting the given lines at the given angles.*

One such point might be as shown here,

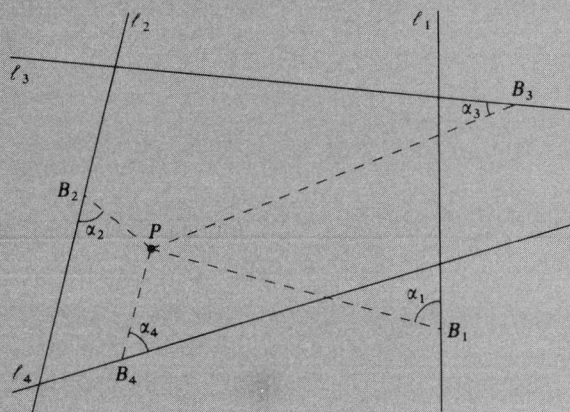

and the locus of all such points is as shown below.

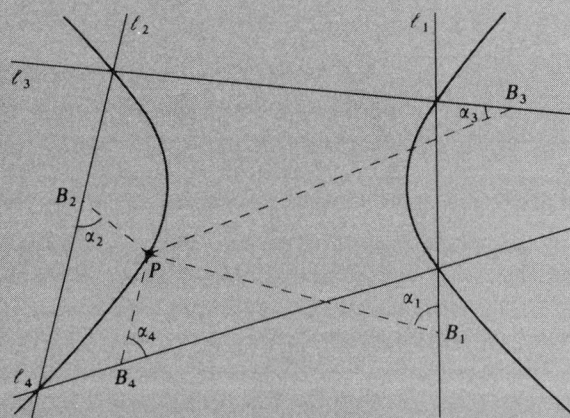

The three line problem is analogous, and can be thought of as the case where two lines of the four line problem are superimposed; that is, there are three lines (ℓ_1, ℓ_2, ℓ_3), three angles ($\alpha_1, \alpha_2, \alpha_3$), and the condition on P is that $p_1 \cdot p_2 = kp_3^2$.

In the account by Pappus (**SB** 11.A4), he first gave the history of the problem, then commented on its solution (in the middle paragraph, which you should find easy to follow, by comparing it with Box 2), then spoke of its further generalisation, to five, six or more lines. Please *read* **SB** 11.A4 *now*.

Question 6

(i)   What did Pappus say was the solution to the problem of three or four lines?

(ii) Did he explain or justify this solution?

(iii) Did he give the solution to the problem of five, six or more lines?

(iv) What difficulties about the latter problem did he discuss?

Comment ——————————————————————————————

(i)   Pappus gave the solution of the 'locus with respect to three and four lines' as a conic section. But he did not go into greater detail about which conic section would arise from which initial configuration of the problem. (That is, for a particular initial specification of the lines and angles, the solution might be hyperbola, as in Box 2, but for another specification it might be an ellipse or parabola, and Pappus did not distinguish these cases.)

(ii) No.

(iii) Pappus did claim that the solution to the problem involving five or more lines would be a curve, so that is part of the way towards reaching a solution. (For it is not impossible, considering the matter in advance, that the locus would turn out to be a collection of points randomly dotted about the plane.) But he went on to say that the curve could not be recognised.

(iv) In attempting to generalise the problem, Pappus noted the difficulties that arose in even specifying the problem, from the dimensional restrictions inherent in Greek geometrical language. If the product of two lines is the rectangle contained by them (that is, an area), and the product of two rectangles is the three-dimensional figure contained by them, and so a volume, then one has run out of dimensions to represent the product of two volumes. This difficulty would arise in the case of more than six lines. Pappus did explain how the difficulty could be got around, by using the language of composition of ratios, but he did not sound very happy about it. ■

There was therefore a clear challenge to future mathematicians, to tidy up and solve this set of problems that Greek geometers had been unable to solve. First, elucidate fully the locus to three and four lines, by describing which conic arises from which initial specification of the problem. Secondly, understand the locus to five, six or more lines.

It was Descartes who took up the challenge, and solved it during five or six weeks of 1633. This is remarkable testimony to the power of his new algebraic analysis. It is impressive that he could quite easily solve Pappus' first problem, the locus to three or four lines. It is even more impressive that he could solve with almost equal ease the locus to five, six, seven or indeed any number of lines. 'I believe that I have . . . completely accomplished what Pappus tells us the ancients sought to do', the jubilant Descartes wrote in *La Geometrie*. His solution is extract **SB** 11.A6 in the Source Book; we shall see that what Descartes understood by a solution—what counted for him as having solved the problem—is an important and revealing aspect of his work. The first part of the extract describes his discovery, in outline, before he goes on to demonstrate the use of his new techniques.

Question 7   Read the first six paragraphs of **SB** 11.A6 (down to where he becomes 'already wearied by so much writing').

(i)   What, in summary, does he claim to have discovered?

(ii)   What is his solution to the Pappus problem going to consist of—does it seem that Descartes will produce a whole curve, or just some points on that curve?

Comment ───

(i)   Descartes had discovered that the solution depends on how many lines there are in the given problem. (Pappus had known, of course, that the case of three or four lines differs from that of five or more, but Descartes managed to distinguish between the higher cases also.) If there are three, four or five lines, the solution can be obtained by ruler and compasses; if there are six to eight lines, the solution can be obtained by conic sections; for ten to twelve lines, the solution needs 'a curve of degree next higher'; and so on. Note that these curves (mentioned in paragraph 2) are not the solution as such, but are the curves needed to produce the solution. The solution itself—that is, the locus being sought—is described in paragraph 4. There he said that for a five to eight line problem, the locus is a 'curve of degree next higher than the conic sections'; for a nine to twelve line problem, the locus is a curve one degree higher still; and so on.

So not only had Descartes found the solutions, and classified them in terms of how many lines were given in the original problem, but also he had made a revealing distinction between the solution—the locus—and what needed to be done in order to construct the solution.

(ii)   Descartes was careful to state that he was looking for points on the locus, not the whole curve itself—the phrase used continually is 'the required points'. This is how it comes about that although a conic section *is* the solution to the three or four line problem, it can be *used for finding* the solution up to the nine line problem.

The distinction between constructing points on a curve, and the curve itself, has important implications for Descartes' conception of curves, as we shall see shortly. ■

So how did Descartes reach these remarkable results, not previously attained by any mathematician of ancient or modern times? His account of that came next. Because of its importance, it is worth spending time teasing out some details of his method.

'First, I suppose the thing done', the opening gambit of paragraph 8, makes clear that this is a method of analysis. The thing that Descartes supposed done is the location of the point C, where C is a point lying on the locus, that is, a point satisfying the property laid down in the problem. If the problem is solved, then there are geometric relations between the lines constituting the final diagram, so that those line lengths are expressible in terms of other line lengths and ratios. In particular, Descartes showed that all the lengths in the problem can be given in terms of just two line segments. It does not seem to matter particularly which segments are chosen; Descartes took the length of one of the lines from C to be y, and the length of some segment along one of the originally given lines to be x. In terms of these two lines, all others in the problem can be written down—that is the content of paragraph 9.

Now it is one thing to express all lines in the problem in terms of x and y, and another to be able to cope with the resulting expressions. But fortunately, Descartes observed (paragraph 10), all the expressions are very straightforward. They are all of the form $ax + by + c$, in effect, where a, b and c are quantities which are known in terms of the problem. For instance, the line CH, the last whose length he deduces in paragraph 9, is

$$\left(\frac{-fg}{z^2}\right)x + \left(\frac{g}{z}\right)y + \left(\frac{fgl}{z^2}\right),$$

which is indeed of the form $ax + by + c$.

The analysis seems to be proceeding very happily, but what has it to do with solving the original problem? It is time for Descartes to invoke the condition laid down in the problem. The locus property which C must satisfy is that the product of certain line lengths, measured from C, is equal or proportional to the product of certain other line lengths. Descartes was now in a position to recast this geometrical locus property in terms of his algebraic analysis.

Question 8   Read the last three paragraphs of **SB** 11.A6 (paragraphs 11–13).

(i)   Summarise the final stages of Descartes' argument that a point on the three or four line locus can be constructed by ruler and compasses.

(ii)   What conception of curve-drawing is implicit in Descartes' procedure?

Comment ――――――――――――――――――――――――――――――

(i)   Using the four line locus as an example, the defining locus property states that the product of two line lengths is equal or proportional to the product of two other line lengths. So each side of this equation is a product of *two* terms of the '$ax + by + c$' form. Such a product can have, when multiplied out, terms in x^2 and y^2 in it at worst, but no higher terms in x and y; it is at most a second degree equation.

Now, the aim of the procedure is to find some point C satisfying the property. The property has now, as a result of the analysis, taken on the form of a second degree equation in x and y. So, Descartes observed (paragraph 13), y can be given any value and the corresponding value of x can be constructed. This is because allocating some value to y leaves a second degree equation in x—that is, an equation with x^2, x and numerical terms at worst—and Descartes had previously shown how x could be found with ruler and compasses in such circumstances.

(ii)   Descartes' conception of finding the locus curve is to find points on the curve, by repeating the procedure of finding the value of x corresponding to different values of y. If one were to do this often enough—an infinite number of times— one would in effect have drawn the curve. The possibility of being able to find, or construct, any point on the curve amounts to knowing the curve, it seems. ■

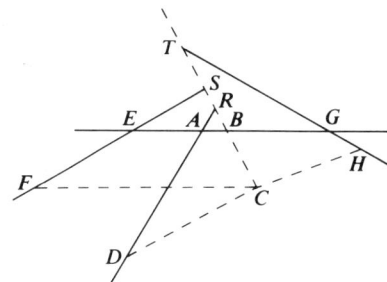

Figure 6

It is worth noting how much Descartes' procedure here differs from the system of 'Cartesian coordinates'. His 'principal lines' are not at right angles (except by accident), and more importantly are not given in advance, but are chosen in relation to the circumstances of the particular problem.

The geometrical construction of algebraic operations, was given at the start of *La Geometrie*.

Descartes' argument here is evidently a quite general one. Given the problem of the locus to n lines, the property will turn out to be expressible as an equation in x and y, of degree about half the number of lines. One can then choose values of y at will, leaving an equation in x of degree $\frac{1}{2}n$ (or thereabouts), whose roots enable points on the locus curve to be pinned down. In summary, the solution to the problem of Pappus is a matter of constructing the roots of equations. This is the perception to which Descartes' algebraic analysis had led him.

How are these roots to be found? The classical mode of construction (by ruler and compasses) enables the root of a quadratic equation to be constructed, and thus the locus problem for three, four or five lines to be solved, by Descartes' procedure. But the locus curve to more lines, whose defining property is a cubic, quartic, or yet higher degree equation, will generally need some construction going beyond ruler and compasses. Descartes needs to convince us, in fact, that a whole array of curves, of ever higher degree, are just as fundamental and constructible as the basic line and circle. If he can show this, then his claim to be able to solve the Pappus problem for any number of lines would be justified, for some allowable curve could be brought in to make possible the construction of the roots of the higher degree equation defining the locus. So Descartes' search for a complete solution to the problem led him to consider what curves could be considered as geometrical. You will see in **SB** 11.A7 the rather remarkable answer he gave. Please *read this now*.

Question 9   What criterion has Descartes given for whether a curve is allowable in geometry? How does the instrument he describes meet this criterion?

Comment ───

We can work out the criterion Descartes was using, from examining the words he used to justify his instrument. He is concerned with *clear* and *distinct* conceptions (as you would expect from an appendix to *Discourse on Method*), and that geometric curves should 'admit of precise and exact measurement'. This specification is, ironically, somewhat imprecise until Descartes spells out what he takes it to mean. We can infer this from understanding what the instrument does.

Although the figure looks rather forbidding (p. 345), the curves which the instrument draws—the dotted lines on the figure, curving out from A—are geometric in the sense that each point is known exactly in its relation to A. The relationship of H (say) to A is, in principle, no more complicated than the relation between two lines in the earlier Pappus problem diagram. The feature which allows these curves to be geometric in Descartes' understanding is that only *one* movement (rotating the bar XY about Y) generates and specifies all subsequent positions of all the points. This is quite different from the double movement that needs to be specified for the spiral, the quadratrix, and other curves that Descartes feels should not be considered as geometric. ■

Descartes' instrument is really one that it is better to hold in the mind than in one's hands—but that, in a way, is Descartes' point. It is a thought-device for showing that curves more complicated than the circle can nevertheless be considered just as accurate, for the purposes of geometric construction, as ruler and compasses. This was not a prospectus for a precision scientific instrument business, but a philosophical investigation into the foundations of geometric truth.

John Aubrey's anecdote of Descartes (**SB** 9.G7) provides another perspective on Descartes' attitude towards mathematical instrumentation.

The question of what curves are allowable in geometry—indeed, what geometrical curves *are*—is a highly important one in *La Geometrie*. For recall that curves have two roles for Descartes: the solutions to problems, and the means by which solutions are constructed. Thus a conic is the solution to the locus problem for three or four lines, as Pappus knew; the locus of points *is* a conic. But a conic is also the *means by which* solutions are found to the locus problem for six to eight lines. The intersection of two conics, or a conic and a straight line, affords the construction of a point on the higher locus. And as the locus curves become more complicated, so it becomes important to know how to recognise a curve that will yield a valid geometrical construction of the solution. Not any curve will do. The quadratrix, for instance, is a nice smooth curve which can be drawn (more or less) with a single sweep of the hand, yet as you saw in **SB** 11.A7 Descartes regarded it as a mechanical curve, not a geometrical one, because two separate motions were involved in its definition.

Descartes' answers to the question of which curves are geometrical are surprising in several respects. On the one hand he wished to rule out certain curves such as the quadratrix and the spiral, because their definitions could not be conceived in a sufficiently clear and distinct way; not even the authority of Archimedes was sufficient to validate the spiral for Descartes. On the other hand, Descartes is quite relaxed about admitting curves constructed by means of string (**SB** 11.A8). Again, he took *pointwise* construction of a curve to be perfectly acceptable; as we saw, he constructed the locus of points solving the Pappus problem point by point, just as a modern microcomputer does. Yet he must have placed some implicit restriction on which curves could be constructed this way, for any curve (even the quadratrix, for instance) can be constructed pointwise if one tries hard enough. In short, the critical distinction of which curves are allowable in geometry does not emerge from Descartes' discussion in anything approaching a clear or distinct way. Most surprising of all, there was a clear and simple answer, implicit in Descartes' work, which he chose not to put forward: to allow into geometry precisely those curves which were represented in his analysis by polynomial equations.

We constructed an ellipse in this way in *Unit 1*, Section 2.

Why did Descartes not take as his criterion of geometric acceptability the property that a curve can be represented by a polynomial equation? This question, among others, was cogently addressed by the historian H. J. M. Bos in 1981; an extract from his paper appears in the source book (**SB** 11.A10). Please *read this now*.

Such curves were to be called *algebraic*, not many years later. This is discussed in the next section.

Question 10 What aspects of Descartes' programme did Bos draw attention to in explanation of the above question?

Comment ───

There were two main reasons for Descartes adopting the approach he did. The more technical one is that for Descartes algebraic curves were not sufficiently geometrical, because they, in turn, would need a rule for their construction before they became intelligible. It was not clear to Descartes (and would indeed have been rather difficult to show) that there was some continuous motion, of the kind he allowed, for the generation of all curves specified in terms of polynomial equations.

The other reason is more general and philosophical. The whole Cartesian programme is an algebraic analysis of *geometry* (as the title of the book suggested); so the grounds of justification must be ultimately geometrical. An algebraic equation was simply not a geometrical object, and could not be taken as the defining property of one without sabotaging his entire enterprise. ■

Even so, it remains true that Descartes' attitude towards the acceptability of algebraic formulations, as seen in *La Geometrie*, was ambivalent. He did not put forward an absolutely clear and consistent position. Bos has traced some of these ambiguities in Descartes' text—reflecting, it is fair to suppose, ambiguities in Descartes' mind—to the early 1630s when Descartes first tackled the Pappus problem:

> There is a conflict in the *Geometrie* between geometrical and algebraic methods of definition and criteria of acceptability. This conflict reflects a break in the development of Descartes' thought about geometry. In an early phase Descartes considered that the aim of geometry was to construct solutions of geometrical problems by means of curves traced by certain instruments; the instruments served as acceptable generalizations of ruler and compass. He tried to find new constructions in this way and to classify them. About 1630 that plan seemed to stagnate and Descartes also became fully aware of the power of algebraic methods. He then changed his programme. Algebra became the dominant tool, both for the solution of problems and for the classification of curves. But Descartes continued to believe in the principle of geometrical construction by means of curves traceable by instruments. As a result, there are conflicting elements in the *Geometrie*.

H. J. M. Bos, 'On the representation of curves in Descartes' *Geometrie*', *Archive for History of Exact Sciences* **24** (1981), p. 298.

So it was a shift in what Descartes was attempting to achieve, in the historical development of his thought, which accounts both for the great significance and also for the internal contradictions in *La Geometrie*. The algebraic language and thinking sits uneasily with the generalised compasses and the solution of problems by finding intersection points of curves.

Recall that, when describing how to draw a curve, Descartes allowed that it was enough to be able to plot arbitrarily many points on it. In this way Descartes gave a geometric interpretation to an algebraic equation in two variables, which was most useful when the equation appeared as a description of the solution to a locus problem. In his example, the equation was a quadratic equation which gave the solution to the Pappus problem for four lines. There is, of course, another way of describing curves, by means of a suitable drawing device. But it is not at all easy to see how to devise a machine in each case. And if you had thought of turning a line of points into a curve by simply joining them up, that could not be claimed to be sufficiently precise. For, as noted earlier, even the quadratrix can be produced in such a fashion. So the original and consistent view of geometrical curves, as precisely those constructible by generalised compasses, was sullied by the algebra which gave Descartes' programme all its power. You might like to re-read the final paragraph of the extract from Bos' paper (**SB** 11.A10) and see what reasons he ascribed to Descartes' decision to go into print with, in fact, a contradictory programme.

There are two main reasons. First, it is easy to understand what it means for curves to intersect, but nothing like so easy to accept that pairs of algebraic equations define geometric points—and Descartes' whole philosophy is based on the centrality of clear and distinct ideas. Second, and perhaps even more important, to define geometric curves in such a way would virtually transform geometry into algebra—and Descartes' aim was to organise geometry, not demolish it. Finally, though perhaps less convincingly, Bos suggests that the great emphasis Descartes placed on finding the simplest solution methods is imperilled by going over to algebra. Indeed it is, but one might say: so what? The interesting fact is that, ever since Descartes' time, some mathematicians have followed an algebraic route, others have gone a more geometric way, and they have differed over precisely this issue. We shall see later that Newton had very firm views on the matter.

Now that we have looked in some detail at Descartes' overall approach, mainly as it appeared in Book I of *La Geometrie*, we can examine rather more briefly what is in the remaining two books. In Book II, Descartes gave a thorough analysis of second degree equations in two unknowns, showing that they represent conic sections, and in particular indicating when the locus in the Pappus problem is a parabola, hyperbola, or ellipse. He then took a special case of the problem of five lines, showed that the equation of the corresponding locus is

$$y^3 - 2ay^2 - a^2y + 2a^3 = axy,$$

and showed how to draw infinitely many points on it by means of a movable parabola and a rotating line. The curve became known as the *Cartesian parabola*, and Descartes was to use it for constructing solutions to equations. Then there followed his method for finding normals to curves, which he regarded as a very important problem indeed. We shall look at this in *Unit 9*, in the context of other methods proposed for finding tangents. He then constructed and analysed several quartic curves (curves of degree four) which he needed in his study of optics. Book II ended with some cryptic remarks about the study of curves in space.

A *normal* is a straight line perpendicular to a curve. Where the normal cuts the curve, it is also at right angles to the curve's tangent at that point.

In Book III he returned in detail to the problem his method would always lead him to: how to solve an equation geometrically. As we saw, a quadratic equation is solvable by circle-and-line construction. Higher degree equations, such as cubics and quartics, require conic sections, and he showed how this could be done. A quadratic equation has a formula expressing its solution, and Descartes showed, as you saw, how to interpret the solution geometrically and thus construct it. A cubic equation also has a solution given by a formula, and Descartes solved such equations geometrically, using a circle and a parabola.

Descartes attributed the formula for a cubic to Scipio del Ferro on the authority of Cardano.

To solve equations of degree five and six requires more complicated curves, which can only be constructed pointwise. Descartes gave but one example, in which solutions to the general equation of degree six were constructed by means of a circle and the Cartesian parabola, before concluding *La Geometrie* with these words, which succinctly capture his ambition and achievement:

> But it is not my purpose to write a large book. I am trying rather to include much in a few words, as will perhaps be inferred from what I have done, if it is considered that, while reducing to a single construction all the problems of one class, I have at the same time given a method of transforming them into an infinity of others, and thus of solving each in an infinite number of ways; that, furthermore, having constructed all plane problems by the cutting of a circle by a straight line, and all solid problems by the cutting of a circle by a parabola; and, finally, all that are but one degree more complex by cutting a circle by a curve but one degree higher than the parabola, it is only necessary to follow the same general method to construct all problems, more and more complex, *ad infinitum*; for in the case of a mathematical progression, whenever the first two or three terms are given, it is easy to find the rest.

> I hope that posterity will judge me kindly, not only as to the things which I have explained, but also as to those which I have intentionally omitted so as to leave to others the pleasure of discovery.

8.4 ALGEBRAIC GEOMETRY— SUCCESSES AND PROBLEMS

We have spent quite some time on Descartes' *La Geometrie* because it influenced the next generation of mathematicians so much. For example, the work of Newton and Leibniz is unthinkable without it. Indeed, as the historian D. T. Whiteside has observed:

> we can without exaggeration say that *Geometrie* was a rich store-house of thoughts awaiting verification and elaboration and extension in the learned commentary. In the half-century after it appeared the study of analytical geometry is largely the history of the improvement and, in some cases, considered rejection of ideas original with Descartes.

D. T. Whiteside, 'Patterns of mathematical thought in the later seventeenth century', *Archive for History of Exact Sciences*, **1** (1961), p. 295.

So Descartes was not the only one to consider algebra and geometry—nor indeed was his study of curves the only concern of *La Geometrie*. We now look more broadly at how mathematics was constituted after Descartes.

Recall that around 1600, mathematicians had felt that rigorous mathematics was synonymous with Greek geometrical reasoning. But they found that rigorous synthetic argument was not conducive to discovery. The original use for algebra in geometry was as part of the process of analysis: breaking the problem down to simpler components until it could be understood. Pappus, and modern commentators like Viète, had stressed the need for following analysis with a synthesis, a putting-together of the small pieces to which analysis had reduced the problem. Algebra was initially supposed to be a general method for reasoning about quantities which themselves are either geometrical (magnitudes) or numerical. Algebraic analysis was initially a discovery method, to discover truths which should then be given a synthetic, geometric proof. So how did the shift develop towards accepting algebraic analysis as a mode of logically convincing proof?

The reception of *La Geometrie* helps to chronicle this shift. It seems that Descartes was dissatisfied with his own work, or perhaps with the difficulty which readers experienced, almost as soon as it was published. Although one doubts if he regretted the frequent injunctions that the reader could (and should) teach himself, he sanctioned an anonymous *Introduction à la Geometrie* in 1638, and welcomed Debeaune's manuscript commentary of 1638–9, *Notes brèves sur la geometrie de Mr. D. C.* But the importance of Descartes' work was recognised in the small circle of European mathematicians, and the energetic Frans van Schooten soon proposed to translate it into Latin.

Florimond Debeaune (1601–1652) was a lawyer who lived at Blois. Descartes felt he understood *La Geometrie* better than anyone.

This translation, when it appeared in 1649, included Debeaune's *Notes brèves* and comments of Schooten's own based on correspondence with Descartes—Cartesian geometry was still not easy. Van Schooten went on to revise this edition during the period 1659–1661, when he added further commentaries by the Dutchmen Johann Hudde, Hendrik van Heuraet, and Jan De Witt, two tracts of his own and another of Debeaune's. Descartes' small work was now escorted by over 800 pages of additional text! Much of this commentary was devoted to the question of finding tangents, and so we shall discuss this later in connection with the history of the calculus (*Unit 9*), but the sheer weight of this response testifies to the importance of Descartes' ideas. Note too the importance, for its dissemination, of translating *La Geometrie* into Latin. By originally writing it in French, Descartes was contributing to the growing tendency to write learned works in the vernacular, but most mathematicians across Europe would still have been able to read Latin more readily than French.

Frans van Schooten (1615–1660) was one of the great mathematics teachers of the period. He lived at Leiden, where he first met Descartes in 1637, during the publication of *La Geometrie* there.

Gradually, other authors took up algebraic geometry. In 1655 the English mathematician John Wallis gave the first systematic treatment of conics in Cartesian terms, and also considered the cubical parabola $y = x^3$ and its intersections with the lines $y + px + q = 0$ for varying p and q. These give the solutions of the cubic equation $x^3 + px + q = 0$, by a method different from Descartes'. In 1659 Sluse published his *Mesolabum*, which discussed cubic equations and, in its second edition of 1668, quartic equations and their geometric solutions by conic sections. In 1679 La Hire published his *Nouveaux elemens des sections coniques* which contained a discussion of Cartesian methods for conic sections, and what is almost a text-book account of geometrical methods for solving equations of various degrees. In it, he discussed Fermat's treatment of the same problem, which had been circulating in manuscript since perhaps the early 1640s, but was published for the first time in that year (in Samuel de Fermat's edition of his father's work). A thorough, indeed prolix, treatment of the solution to all cubic and quartic equations by intersecting circles and parabolas was given by the English clergyman Thomas Baker in *The Geometrical Key* (1684), which became widely quoted because its answers were so explicit.

René-François de Sluse (1622–1685) was a canon of Liège cathedral.

Philippe de La Hire's work on projective geometry was discussed in *Unit 7*, Section 3.

The bulk of these treatments concentrated on showng how conic sections can be used to solve polynomial equations of higher degree, but they inevitably led their authors to study higher degree curves in their own right. The simplest case, cubic curves, was already very complicated, and undoubtedly the greatest single accomplishment in this aspect of algebraic geometry was Newton's astonishing classification of cubics in the later 1660s, although this work was not published until 1707, as an appendix to his *Opticks*.

We looked at this aspect of Newton's work in *Unit 7*, Section 3.

What are we to make of these developments? In a sense, there seems to have been a profusion of activity, but in another sense there was only gradual progress. We must not forget, of course, the relatively small number of mathematicians active during the seventeenth century, whose attentions were caught also by other burgeoning topics, notably the calculus. It will help to collect our ideas under three headings.

1   Conic sections
2   The theory of equations
3   Higher plane curves—those lying in a plane, other than straight lines and conics.

1   Conic sections

Apollonius' truly monumental classical treatment of the conic sections had been available since 1566, in the translation of Books I–IV by Commandino. This made it easy for mathematicians to know what was true; and easy for them to know what to re-express in Cartesian language if they saw fit to do so. The question is: did they so choose?

Question 11   Before reading on, can you suggest in the light of what you have read, any reasons for mathematicians to treat conic sections algebraically?

Comment ──────────────────────────────────
One reason might be pedagogical—recall how difficult Apollonius' *Conics* was agreed to be. Another reason might be to make the theory of conics more readily applicable to other problems, for example to problems in the theory of equations.

Perhaps you thought of other possible reasons; if you did, bear them in mind as we proceed. ■

In fact, the study of conic sections was only gradually made algebraic. For example, Mydorge's *Coniques* (1639) was thoroughly classical, although on occasion more elegant than Apollonius. And Desargues' profound, if unsuccessful, reformulation of their study was entirely geometrical. But in 1655 there appeared the earliest simple account of Conics which derived from the Cartesian approach, John Wallis' *De Sectionibus Conicis*. Praising the proofs of symbolic arithmetic for their brevity and perspicacity, Wallis showed how to obtain equations for conic sections with respect to perpendicular axes chosen in the plane of section, so that he could then consider the curves 'without regard to their customary origin'. So Wallis eliminated the need to deal with three dimensions, cones, and sections, by deriving equations for the sections and also by showing how their most basic properties (diameters and tangents) could be determined from the equations alone. To derive the equations was a trivial task to anyone familiar with Apollonius, but to solve geometrical problems using them alone was not. However, the conics were still defined as sections of a cone initially. The much more radical step of starting instead with second degree equations, which were then interpreted geometrically, was not to be taken decisively until 1748.

This step was taken by Leonhard Euler, and is discussed in *Unit 12*.

An interesting insight into mathematicians' reception of different ways of treating conic sections is afforded by the work of Philippe de La Hire. Recall from the last unit that La Hire had presented a projective treatment of conics in a book of 1673. If you look at **SB** 11.B2 you will see how he judged it had been received, and how he responded six years on.

It is clear from this that he decided his mistake had been to deal so boldly with the subject as a three-dimensional one, for not all readers seem to have shared his enthusiasm. The new treatment (*Nouveaux elemens des sections coniques* (1679)) fared much better, because it was entirely two-dimensional. The historian J. L. Coolidge has written of it that:

> It would be hard to find a book offering an easier introduction to the conics. Each type of curve is considered separately, starting with some characteristic property. The ellipse appears as the curve where the sum of the distances from the two foci is constant. This leads immediately to the properties of the tangent, the conjugate diameters, and the equation in Greek form. The book would be as usable today as it was the day it was written.

J. L. Coolidge, *A History of The Conic Sections and Quadric Surfaces* (Oxford University Press, 1945) p. 41.

But it is interesting to note that La Hire avoided the use of Cartesian methods in this part of his work. What Coolidge meant by 'the equation in Greek form' is an equation between magnitudes written out as Apollonius had done.

We called this the *symptom* of the curve in *Unit 4*.

La Hire reserved his explanation of Cartesian methods for his discussion of higher plane curves.

Question 12 In contrast to your earlier answer, suggest why mathematicians might *not* have thought it desirable to present the theory of conic sections algebraically. (One reason for asking this question is because mathematicians were genuinely divided on the matter.) Can you give a pedagogic reason as well as a more philosophical one which might underlie a reluctance to adopt a Cartesian approach to conics?

Comment —————————————————————————

The pedagogic virtue of a purely geometrical presentation was that it described the theory of conic sections in the same language in which elementary geometry was learned and taught, without having first to present the new and complicated techniques introduced by Descartes. Another—philosophical, if you like—virtue was that the analysis of geometrical problems is somehow more appropriately done in geometrical terms (lengths and areas) than in algebraic terms, alien to the subject. To understand such a proof was to understand an argument about lines and shapes, not a process of manipulating letters. ■

This point was to be put with surprising vigour by Newton as late as 1707, as you will see. Since Newton's response to Descartes is worth a section to itself (following this one), here we shall illustrate the spread of views with two other writers. John Craig was a young Scottish mathematician and friend of Newton; his *Nova Methodus* (1693) adopted Cartesian methods and started from the idea of equations of the second degree, which he showed describe curves which are identical with conic sections. On the other hand, Guillaume de L'Hôpital wrote a successful *Traité analytique des sections coniques*, published in 1707 (after his death), which used Cartesian coordinates but otherwise defined the conics by reference to their focal properties much as La Hire had done. So he was more inclined to favour geometry.

The tension between what today is called the rigour of algebra and the intuitive force of geometry is a permanent part of the history of the subject. You will meet it again in later units, on geometry in the eighteenth and nineteenth centuries. It derives from a feeling that shapes are prior to symbols, and that mathematics is about curves not letters. So ultimately a theorem is about two curves meeting, for example, and not two equations having a common root. Belief in the primacy of geometry also derives from a feeling that sometimes you can almost literally see why a geometric proof is true, whereas an algebraic proof does not have such a power to convince. It is countered by the argument that mathematics *can* be about letters if mathematicians so choose. Further, the much-vaunted visual aspect of geometry is something that not every mathematician can share, and indeed can be downright misleading. Perhaps, said algebraists, the practice of mathematics is telling us to change our ways.

Note, too, how the rigour and the intuitive qualities have swapped places!

Happily, the question 'geometry or algebra?' is not resolved. The existence of a well-developed and truly geometric theory of conics enabled geometers in the seventeenth century to use Cartesian methods selectively according to their preferences. It was not to be so easy once curves more complicated than conics came to be considered.

2 The theory of equations

We have seen that Descartes devoted much effort to showing how simple curves can be used to solve equations. In particular, he constructed the solution to every type of cubic equation by using a circle and a parabola, and proved that every quartic equation could be treated in the same way. But the detailed working out was far from explicit, and much remained for later writers to clarify. René-François de Sluse, for instance, dealt with the solutions of third- and fourth-degree equations in the 1668 edition of his *Mesolabum*. Below is Sluse's technique for solving the cubic equation $x^3 = ax + b$.

Sluse converted this to a proportion between two ratios—arguably an old-fashioned thing to do in 1668:

$$a : x^2 = x : x + b/a.$$

You can check that this is correct by cross-multiplying.

To bring in known curves, he introduced a second variable y as the mean proportional between x and $x + b/a$,

$$x : y = y : x + b/a,$$

whence $\sqrt{a} : x = x : y = y : x + b/a.$

He had now, in effect, arranged matters so that x and y were the two mean proportionals between $\sqrt{a}$ and $x + b/a$. So, by the classical argument of Menaechmus (*Unit 3*, Section 3, Box 4), he had three conic sections: the parabola $x^2 = \sqrt{a}\, y$, the hyperbola $y^2 = x(x + b/a)$, and the hyperbola $xy = (x + b/a)\sqrt{a}$, any two of which provided him with a construction for the roots of the original equation. So Sluse was drawing upon both the Greek geometric heritage and the newer Cartesian methods in devising his solution. The balance was if anything tilted towards the distant past, as the very title of his book makes clear.

The *Mesolabum* was an instrument for constructing mean proportionals, such as that devised by Eratosthenes (**SB** 2.F3).

But, as Descartes had discovered, conics alone will not enable you to solve equations of degree higher than four. What is more, in the process of trying to solve an equation geometrically, mathematicians conducted the following kind of analysis.

> Given an equation to be solved, find two curves whose intersections have the roots of the given equation for their x-coordinates.

This was worth doing only if the found equations are of lower degree than the given one. Generally they used two curves of about the same degree. So if a quartic had to be solved, two curves of degree two would suffice—but to find their equations involved a lot of algebra. Both Sluse and Baker, as mentioned earlier, wrote books on just this problem. This was certainly a stimulus to put the theory of conics into algebraic dress. It is therefore interesting to see that La Hire introduced Cartesian methods into the study of curves in general, in the second part of his *Nouveaux Elemens* and then used curves expressed in this form for the construction (geometrical solution) of equations. Most of the curves he used in fact turned out to be conics—still the easiest curves to treat—but the stimulus to study curves of higher degree remained.

For example, if the equation to be solved was of degree n, the degree of both solving curves would be k, where $k^2 \geqslant n$.

SB 11.B3

3 Higher plane curves

For plane curves other than conics, the situation is entirely different. Without the language of Cartesian equations almost none were discovered. But once such methods were available a profusion of cubics and quartics appeared and were given wonderful names, like the *witch of Agnesi*, the *folium of Descartes*, the *trident of Descartes*, and so on. In 1684 Leibniz proposed that a distinction be made between *algebraic* curves, which are defined by polynomials in x and y, and *transcendental* curves, which are not. With this distinction comes the idea of a general theory of curves based upon algebraic definitions, and here we touch again on a significant problem. The question is: what is a curve? Or, more precisely, what kind of curves can be studied mathematically?

In this unit we have seen several kinds of answer to this question. A plausible answer at the time would have been some restriction on 'anything drawn by a free motion of the hand', to use an eighteenth-century phrase for curves. This is too vague to be much use without more clarification, because it immediately admits so many curves about which nothing is known, so a second answer was: curves generated by a suitable machine. We saw in Section 3 of this unit how Descartes struggled to limit curves drawn by continuous motion to a small class his algebraic methods could deal with. But we also saw there why he could not simply define that class as the class of algebraic curves, curves defined by polynomial equations. Leibniz' viewpoint shows that he had no scruples about accepting both the class of curves defined by polynomial equations, and also curves defined in other ways. In place of Descartes' obscure distinction between the curves he would accept into mathematics and those, like the quadratrix, which he would not, Leibniz put the much clearer distinction between curves which have algebraic equations and those which do not. In an important shift of emphasis, he accepted both sorts, provided they could be defined precisely.

That said, how much of a theory of higher plane curves, or even of algebraic curves, was developed in the later seventeenth century? Work in this area is closely related to the history of the calculus, for the curves were often studied as proving grounds for new techniques of quadrature or tangent-finding. Thus Christiaan Huygens, in the 1650s, investigated the cissoid and the conchoid, two curves that had come down from the Greeks, but this was to practise his grasp of Descartes' method of tangents. Or again, Fermat studied the curve $y = k/x^2$, but this was to demonstrate his method for finding areas.

SB 11.C5

It often turned out, however, that new techniques enabled new properties of the curves to be found. For example, Fermat showed that the area bounded by the x-axis, the vertical line $x = 1$, and the curve $y = k/x^2$ is finite, even though its boundary is infinitely long—a rather curious fact. Overall, the study of higher plane curves—with the exception of Newton's remarkable study of cubics—was inseparable from that of the pre-calculus and the calculus itself.

Unit 7, Section 3

By defying classical methods but yielding to the (post-) Cartesian ones, the higher plane curves taught mathematicians to use and trust algebra. Conics could remain quite classical, but higher plane curves required algebra. Forced to rely on symbols even to do geometry, a tendency the demands of the calculus also emphasised, mathematicians were being driven from the garden of geometry. In the eighteenth century they left eagerly for the new lands of algebra. And no-one felt the tension more keenly at the end of the seventeenth century than Newton.

8.5 NEWTON AND DESCARTES

The one geometer who could claim to be Descartes' peer in the seventeenth century, the one to whom Descartes was most frequently compared by his successors, was Isaac Newton. Not only did Newton rival Descartes with his breadth of interests, but also much of his work can profitably be seen as a rich and almost personal struggle with his Cartesian inheritance. So although Newton has already flitted briefly across these pages, and will appear more substantially when we deal with two of his major achievements (the calculus in *Unit 9*, and his *Principia* in *Unit 11*), we take this opportunity to become better acquainted with him.

Newton was born on Christmas Day 1642 in the manor house of Woolsthorpe near Grantham in Lincolnshire. He was a sickly child, and his father was already dead. When his mother remarried he was sent, at the age of three, to live with his grandmother, and he grew up with this old woman whom he never came to like, much less love. At the age of 10 he was briefly re-united with his mother before being sent away to live in Grantham, where he attended the Grammar School.

Figure 7 Isaac Newton (1642–1727)

By 1661, when he entered Cambridge University, he was already confirmed in certain traits. He would immerse himself in thought to the point of neglecting his meals; he was estranged from others and kept his own company; his mind was never at rest. It was always to be so. At Cambridge, the syllabus was undemanding and did not reflect the intellectual tumult of the age. But it left the young man, already eager to learn and explore, free to follow up what clues he found. He kept a notebook, and in 1664 began to fill it with a rush of observations; both practical experiments and the fruits of his reading were entered. Many notes are on Descartes: his theory of light, his theory of planetary motion, his theory of the tides. Other passages mark the beginning of Newton's interest in mechanics, and it seems he first read Keplerian astronomy at this time. Yet others reflect his absorption in the philosophical ideas of Henry More, the leading Cambridge Platonist of the day and an influential writer. More's central concern was to reconcile mechanical philosophy with Christian religion by resolving this conundrum: how can God act in a world which obeys physical laws? His analyses of this problem impressed themselves upon the devout Newton, and may well have helped him towards his own novel ideas of universal gravitation.

Not least, Newton discovered mathematics. Looking back 35 years later, in 1699, he tells us that,

> By consulting an accompt of my expenses at Cambridge in the years 1663 and 1664 I find that in ye year 1664 a little before Christmas I being then senior Sophister, I bought Schooten's Miscellanies & Cartes's Geometry (having read this Geometry & Oughtred's Clavis above half a year before) & borrowed Wallis's works and by consequence made these Annotations out of Schooten & Wallis in winter between the years 1664 and 1665. At wch time I found the method of Infinite series. And in summer 1665 being forced from Cambridge by the Plague I computed ye area of ye Hyperbola at Boothby in Lincolnshire to two & fifty figures by the same method.

Newton's early, self-taught, struggles with Descartes were recorded at the end of his life:

> He bought Descartes's Geometry & read it by himself when he was got over 2 or 3 pages he could understand no farther than he began again & got 3 or 4 pages farther till he came to another difficult place, than he began again & advanced farther & continued so doing till he made himself Master of the whole without having the least light or instruction from any body.

'Miscellanies' is Book V of van Schooten's *Exercitationes Mathematicae* (1657). 'Cartes's Geometry' was van Schooten's 1659–1661 edition of Descartes' work.

D. T. Whiteside (ed.) *The Mathematical Papers of Isaac Newton*, vol. I (Cambridge University Press, 1967) pp. 7–8.

John Conduitt, 'Memorandums relating to Sr Isaac Newton given me by Mr Abraham Demoivre in Novr 1727', cited in R. S. Westfall, *Never at Rest* (Cambridge University Press, 1980) pp. 98–99.

For once, Cambridge may deserve some slight credit, for although Isaac Barrow, the first Lucasian professor of mathematics, was never Newton's tutor, it is hard to see who else could have stimulated Newton's interest in mathematics, or lent him a copy of Wallis. That said, this strange young man proceeded on his own, as was always his fashion, and in the eighteen months to Spring 1666 thought of nothing else but mathematics. His invention of the calculus which resulted will occupy us considerably, later on. He turned again to mechanics to grapple, unsuccessfully, with the motion of the planets. What caused these large and fast-moving objects to stay in orbit and not rush away from the sun? The legend of the apple refers to this period (if not literally it is poetically true, an almost Biblical apple) and suggests that Newton, even then, sought to extend known, terrestrial mechanisms to celestial problems. Then came the dedicated explorations of optics and the nature of colour and light, which for many years were Newton's most public demonstrations of his genius.

In 1669, when Barrow resigned his chair to pursue his chosen career as a divine, Newton was appointed his successor—and, given the highly political atmosphere of Cambridge, Barrow presumably had a hand in it; there is an oft-told story that he resigned in Newton's favour which is unlikely. In any event, Newton was now financially secure and free to devote himself to research. He also had to lecture, by the Statutes governing his chair, on 'some part of Geometry, Astronomy, Geography, Optics, Statics, or some other Mathematical discipline'.

Cited in Westfall, *Never at Rest*, p. 208.

This does not seem to have been a great success. It was recorded a few years later that:

> so few went to hear Him, and fewer y^t understood him, y^t oftimes he did in a manner, for want of Hearers, read to ye Walls.

Westfall, *Never at Rest*, p. 209.

Nor did his individual teaching thrive any better. Newton seems during his whole time at Cambridge (where he was Lucasian Professor until 1696) to have had only three serious students. Here, embarked and 'Voyaging', as Wordsworth was to put it 'through strange seas of Thought, alone' we may for the while leave him. You may glimpse more of his progress in the extracts from J. M. Keynes' provocative talk (**SB** 12.F4) which initiated the modern biographical study of Newton. We now look at the question of his complicated relationship to the work of René Descartes.

SB 12.E3

Newton took his algebraic symbolism from Descartes. He wrote equations to describe curves, as Descartes had done. He cared, as Descartes had, for establishing the priority of the geometric over the algebraic. And by his achievement he raised the creative tension we have traced in *La Geometrie* to a state of complete incoherence. The result was a palace.

'At which time' he wrote of the winter of 1664–5 'I found the method of *Infinite series*'. Harmless-sounding words, but they represented an enormous increase in the power of mathematics. Indeed, circumstances of publication (to which the solitary Newton was averse) meant that for many years Newton's reputation as a mathematician rested on his prowess with infinite series. Leibniz, for example, was to regard Newton as a first-rate mathematician on such grounds, without knowing that he had already invented the calculus. So what are infinite series, and what did Newton contribute?

All of Descartes' algebra involved finite expressions (*polynomials*). Every term and every equation in his work can be written down completely. An *infinite series* is typified by this one, which is one of the first Newton was to write:

$$x - \tfrac{1}{3} \cdot \tfrac{1}{2}x^3 - \tfrac{1}{5} \cdot \tfrac{1}{8}x^5 - \tfrac{1}{7} \cdot \tfrac{1}{16}x^7 - \tfrac{1}{9} \cdot \tfrac{5}{128}x^9 \text{ etc.}$$

Something like a polynomial, then, but going on for ever in a way that can only be described by a hand-waving gesture, or 'etc.', or dots (. . .). We receive a sense of their importance from a fascinating document, Newton's second letter to Leibniz or *Epistola posterior*, of 24 October 1676.

SB 12.C2 Strictly, the letter was written to Henry Oldenberg, secretary of The Royal Society, for copying and transmission to Leibniz. Oldenberg acted as a coordinator of scientific communication, somewhat as Mersenne had done.

This is a famous source documenting Newton's earliest work, but it is not completely self-explanatory. Newton had been reading Wallis' *Arithmetica infinitorum* where an ingenious *interpolation* method, as Wallis called it, was explained for finding the quadratures of certain curves. If, for instance, the areas under the curves $y = 1$, $y = 1 - x^2$, $y = (1 - x^2)^2$, and so on, are known—as they were, by this time—then the areas under curves with the more awkward equations

$$y = \sqrt{(1 - x^2)}, \quad y = (1 - x^2)\sqrt{(1 - x^2)}, \quad \text{etc.,}$$

can be calculated. This is done by noticing that all these equations form a patterned sequence, when expressed in exponent notation, in which known and unknown quadratures alternate: the right hand sides are

$$(1 - x^2)^{0/2}, \quad (1 - x^2)^{1/2}, \quad (1 - x^2)^{2/2}, \quad (1 - x^2)^{3/2}, \quad (1 - x^2)^{4/2}, \ldots$$

So the quadrature of $y = \sqrt{(1 - x^2)}$—that is of the circle $x^2 + y^2 = 1$—must lie, by some principle of continuity, between the quadratures of $y = 1$ and $y = 1 - x^2$. Newton described, in the first part of the *Epistola posterior*, how he was following through Wallis' argument when he discovered that the quadratures of the awkward curves could be expressed as infinite series. He seems to have done this by a process of shrewd guesswork and pattern recognition. The significant leap of the imagination was *to accept an infinite series as the answer*. This is going far beyond what Descartes would have allowed as acceptable.

A few weeks later, Newton looked again at his notes, and saw both how to simplify his work, eliminating the need for creative guesswork, and how to progress further. This he described in the rest of the extract from the *Epistola posterior*. Please read that now (**SB** 12.C2(a)).

Ignore the algebraic calculation displayed at the end of the extract, which is more complicated than we need to go into.

Question 13   In the second half of the extract (starting from 'But when I had learnt this,') what does Newton claim to have discovered, and how does he prove it?

Comment

Newton saw that his discovery had nothing essentially to do with quadrature-finding, but was about the expression, as infinite series, of things with fractional exponents. So he applied the same technique as before, to the curve equations themselves, such as

$$y = (1 - x^2)^{1/2},$$

and came up in this case with the infinite series

$$1 - \tfrac{1}{2}x^2 - \tfrac{1}{8}x^4 - \tfrac{1}{16}x^6 \text{ etc.}$$

Thus he had found a rule for 'the general reduction of radicals into infinite series'.

In order to check that this was right, he squared the infinite series, multiplying it by itself term by term, and found that indeed he got back to $1 - x^2$, and similarly for the others. Finally, Newton re-extracted the square root arithmetically, to show that the initial geometrical context was not essential. ∎

Newton did not rest here. He now knew that he had a general rule for handling expressions like $(1 + x)^{m/n}$. They could be written as infinite series, and those series treated just like polynomials. You will see later how powerful this technique is when coupled to the calculus, but it is quite striking on its own, for it brings *transcendental* curves into the orbit of algebraic analysis. A transcendental curve, such as the cycloid or the quadratrix, cannot be described by a polynomial equation but it can be handled by infinite series. Descartes' abhorrence of these curves was not shared by Newton, not least because they were, thanks to infinite series, amenable to algebraic analysis.

While still in the first phase of his voyage through the frontier of mathematics, Newton surpassed Descartes in another way. This was his complete enumeration of cubic curves, made during the late 1660s, and his first study of their geometrical properties. This was mentioned on pp. 22–3 of the last unit, and you might like to look back now at those pages and at Newton's own drawings of some of the curves (in **SB** 12.D2). We see how completely Newton was indeed the 'master' of geometry and of *La Geometrie*, for this work is a Herculean feat of symbol manipulation. However much we may want to suppress the fact from history and mathematics courses alike, the sad truth is that success in mathematics is often a matter of calculating lots of interesting examples and being lucky. Newton would not have been Newton had he not been willing to calculate for weeks on end, but he also lived at a propitious time.

But there is another point. Where does geometry stop, and algebra take over? Much later, in the 1680s and 1690s, Newton took to writing about this question. We have manuscripts of lectures, and a book, the *Arithmetica Universalis* (published in 1707, but written much earlier) in which he put down his thoughts on the matter. Please *read* **SB** 12.D3 *now*.

Question 14 What, in Newton's view, is the criterion of geometrical simplicity? How does it bear on the distinction between algebraic and transcendental curves, and on Descartes' classification of curves?

Comment ——————————————————————————————

Simplicity is ease of generation, so the circle is, as the Greeks proposed, simpler than the parabola, and the cycloid is simpler than any algebraic curve of high degree. So he does not observe a distinction between algebraic and transcendental, and he completely disagrees with Descartes' views on curve classification. ■

We earlier suggested Newton's ideas formed a palace of incoherence. It is impossible to extract from their richness, without injustice, a coherent philosophy. On occasion, geometrical simplicity commended itself to Newton; at other times his working practice was an orgy of calculation. Newton was so impressed by his own power of thought (not unjustifiably) that he had great difficulty in acknowledging the contributions of others—in his account of infinite series, for example, he greatly played down the significant input of Wallis' ideas. Nor was Newton above administering pontifical rebukes. If you look at **SB** 12.D1 you will see a delightful comedy.

It was only in the 1680s that Newton started to fill in the gaps in his education—you have seen how he was almost self-taught. So he came late to the study of the Greeks. In this extract he applied himself to the resolution of the Pappus problem, the locus to three or four lines. Newton solved it as any (gifted) Greek could have done, and on those grounds impugned Descartes for making 'a great show as if he had achieved something'. (Not that we should feel too sorry for Descartes, who yielded nothing to Newton in his need to assert that he was mathematically beholden to no-one else.) Should we be grateful that Descartes missed this solution to the Pappus problem, or did he know that his algebraic methods were immensely more general than any classical method could be? Newton's riposte is curiously inappropriate for the end of the seventeenth century.

If it is right to see a struggle in the mind of the fatherless Newton to live with the discoveries of Descartes—and not all historians would pursue such a line of analysis—the central intellectual disagreement must be over the role and concept of motion, and that in two ways. First, to Newton curves were often generated by motion, in a more vivid, less abstract way than in Descartes' conception. Unlike Descartes, Newton allowed transcendental curves, and above all exploited the idea of motion for finding tangents to curves, where Descartes was more statically algebraic (as you will see in *Unit 9*). Second, Descartes had a theory of planetary motion which was widely accepted. It was this theory that Newton destroyed in *Principia*, and while the details can wait until *Unit 11*, it is worth noticing now that this truly substantial disagreement with Cartesian cosmology occupied Newton in what was, arguably, the most important work of his life.

It is fitting to end by contemplating Newton's own assessment of his relationship to Descartes. He wrote to Hooke that,

> If I have seen farther than Descartes, it is because I have stood on the shoulders of giants.

Should one insist on the 'if' with its suggestion that he has not? Should one suspect insincerity, on the grounds that there were no giants? And is it a compliment or an act of denigration to single out Descartes as the man you have surpassed? Newton knew as well as any one that there was no greater in his chosen fields—the comparison was as inevitable as it is instructive.

Gaukroger, Stephen (ed.), *Descartes: Philosophy, Mathematics and Physics* (Harvester Press, 1980).
> This is an interesting set of essays and includes one by Mahoney which was discussed in Section 1 of this unit.

Rée, Jonathan, *Descartes* (Allen Lane, 1974).
> There is a huge literature about Descartes, but this book is a bold attempt to give an overview in 200 pages. Covers much of his philosophical, as well as his scientific and mathematical, work.

Smith, D. E., and M. L. Latham (trs) *The Geometry of René Descartes*, (Dover Books, 1954).
> This is a parallel text with an English version facing a facsimile of the French original. Most of Descartes' philosophical works have likewise been translated into English, in many editions.

Sorell, Tom, *Descartes* (Oxford University Press, 1987).
> A good and accessible intellectual biography of Descartes in an inexpensive paperback, giving judicious consideration to the position of mathematics in Descartes' work as a whole.

Suggested reading relating to Newton will be found in the Further Reading section of *Unit 9*.